© 2013 Daniel Rhys Ford, IV
All rights reserved
No part of this publication may be
reproduced in any form by any means
known now and of technology known
in the future without written permission
of the publisher and the author.

Copyright Registration number
Txu-1-777-575 September 2011

ISBN-13: 978-0-9855295-2-9
Library of Congress Control Number 2013933489

© 2013
The Marlin & The Mermaid "Help Save the Chesapeake Bay"
All rights reserved

Dedication

To our children, the future leaders and policy makers for our society,
for they will "Help Save the Chesapeake Bay"

To the Marlin and the Mermaid in my life;
Rhys and Olivia.

The Marlin and the Mermaid
"Help Save the Chesapeake Bay"

Based on characters created by Daniel R. Ford

Written by Daniel R. Ford

Rick and Daughter Camie Romano - Cover and interior illustrations

Pete Crossley - Graphics

Published by Green Seahorse Media LLC

About the Author - Daniel R. Ford

I love H_2O!

In 1969 our family moved near to the beach. Growing up around the ocean I had a unique opportunity to experience what a precious commodity we have in this huge natural resource.

Learning to fish and surf provided an experience for some of the most amazing natural scenes. Later in life I became a commercial crabber, then joined the U.S. Merchant Marines and traveled around the world.

With this personal knowledge gained, I have the opportunity to write an educational story about our waterways, how we can make better decisions to keep them clean, and instill these values with our children.

About the Illustrators - Rick & Camie Romano

Rick is a Virginia Beach artist who has made a living drawing and painting in the saltwater environment he's grown up with. An avid windsurfer and surfer for over twenty years, Rick is forever inspired by the timeless beauty and power of the sea. His unique style colorfully reflects his passion for the ocean worldwide.

In his earlier years, many of his works were of coastal landmarks such as old beach cottages, lighthouses and popular beach locations. Today with the growing demand for tropical and surf-related themes, Rick is devoting his studio time for exactly these themes. Camie Romano, Rick's daughter, is attending college in Virginia and contributed her artistic talents to this project bringing to life the mermaid and other characters. This is the second awesome book they have illustrated together!

In the heart of the ocean
in the deep blue sea,

lived a beautiful creature
and a hero to be.

One was a Mermaid with radiant skin,

dark emerald eyes, and a curvy tail fin.

She knew all the creatures
like the back of her hand,

she liked to relax
on the white fluffy sand.

She was fast as a Wahoo,
strong as a Whale,

this giant Seahorse had
a cute curly tail!

The hero was a Marlin,
strong and fearless was he.

He swam all the oceans
and all seven seas.

He was the largest Blue Marlin
of his class,
and one thing he didn't like,
he didn't like trash!

Here is the "tail"
of the fabulous two,
how they want to protect the oceans
for me and for you!

Their story started,
one bright sunny day,
as the Marlin and the Mermaid swam
in the mouth of the Chesapeake Bay.

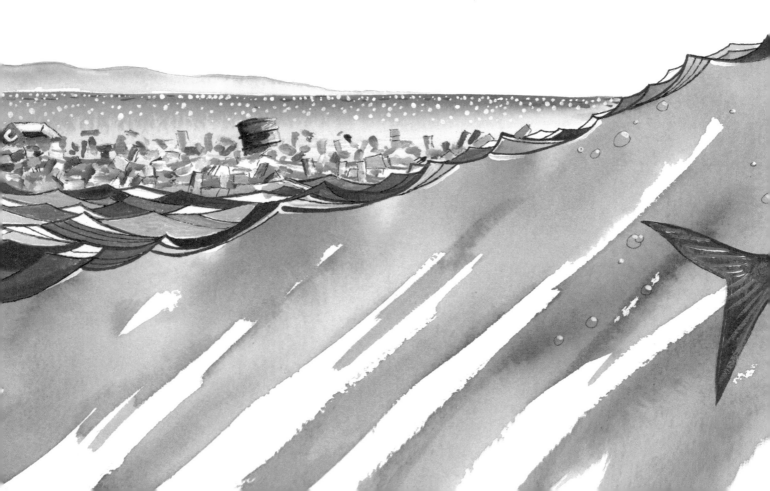

When off in the distance
they saw a trail of debris,
slowly drifting
right on out to sea!

They were amazed at the trash they had seen,

they wondered why the humans can't keep their land clean.

So they started on a mission
to find the source of the cause,
to try and put
an end to it all!

They followed the trail of trash
into the bay they swam.
Where they met a little fish
who said; "Guess who I am?"

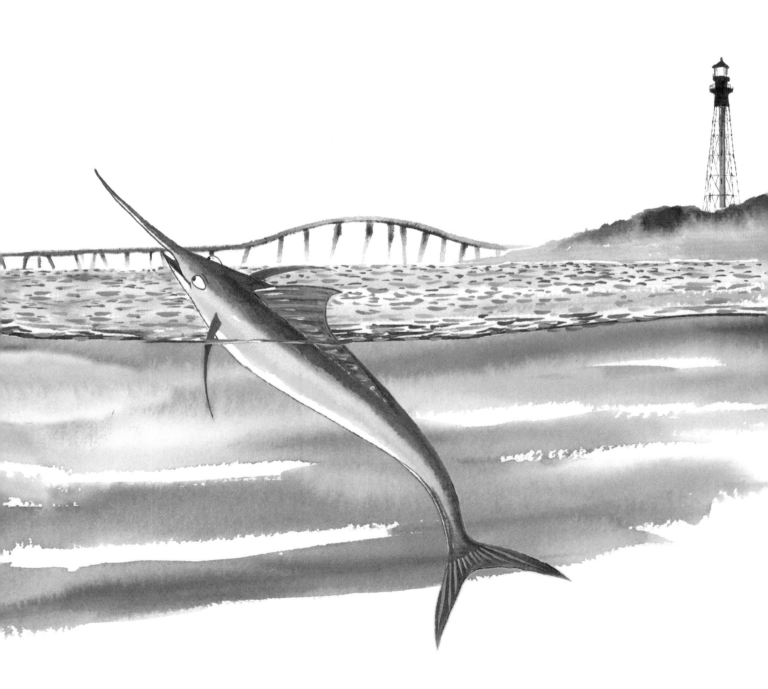

"I'm small, I'm oily,
by some considered a treat,
by the bigger fish
I occasionally get to meet."

I swim all night,
I swim all day,
I filter water
for the Bay.

I swim in schools
too large to count,
it's pretty hard
for predators to surmount!

Following the path of trash,
this took them to a sea-side town.
Where along the salty marsh
they spotted something brown.

As they took a closer look
they peered into a bed,
they were marveled and amazed,
as they scratched their head.

For tucked into
a little cove,
were thousands of Oysters
by the drove.

There the Oysters raised their voice
and this is what they said;

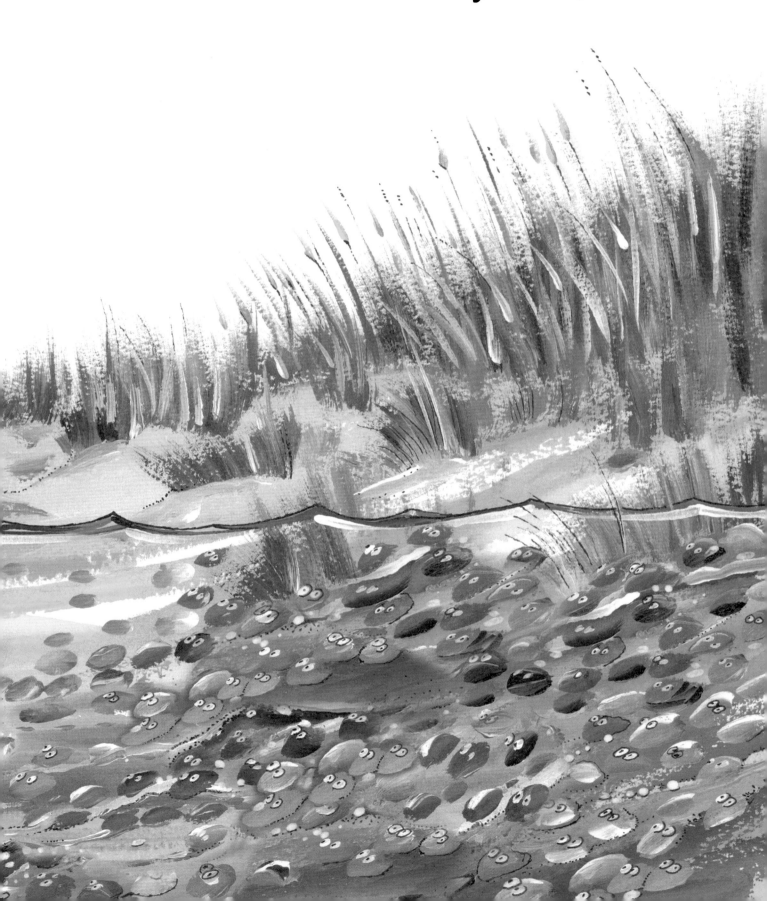

"We remove Algae from the Bay,
filtering seawater two gallons every hour,
joined by thousands more
we have real power!

For once we could
filter the entire bay,
not in months, or weeks,
but in three days!"

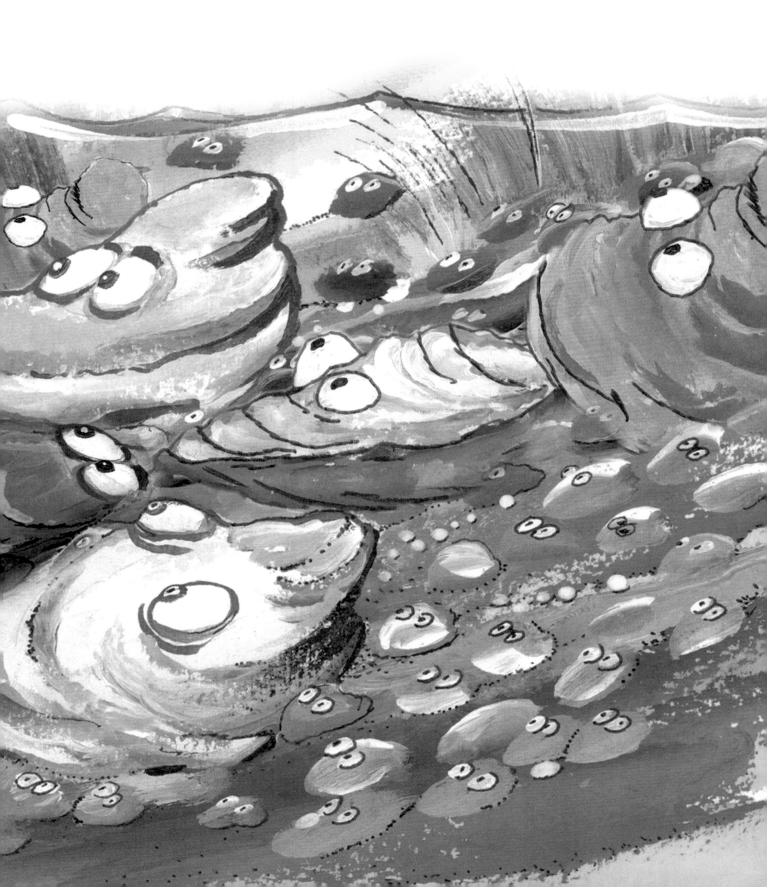

But now our numbers
are so low,
we need to find
somewhere to grow.

If we can find
a safe place to stay,
We'll make a big difference
To "help save the Chesapeake bay"

Then they said good-bye to their dear Oyster friends!

The Marlin and the Mermaid continued their quest, dodging boats, Sharks, and stinging Jellyfish.

When out of the corner of their eye, they saw something red.
In this tide
all the fish were dead!

A toxic Algae bloom
was sweeping the Bay,

killing everything that couldn't
get out of the way!

An Algae bloom
as you can see,
takes all the oxygen
right out of the sea.

Starving all the creatures
both great and both small.
It is caused by nutrients
and fertilizers that can harm us all!

So they did
what they could do,
they warned the Minnow
and they warned you!

Stay away from this thing
they call "red-tide",
until it moves on
and passes on by!

As the trash became thicker,
with more debris found that day,
they started up the inlets,
straight out of the Chesapeake Bay.

Up on the grass,
lying next to a drain,
was all of the trash
just waiting for a rain!

Finally the Marlin and the Mermaid found the source of the trash. They had one question they decided to ask.

What could you do
to put to an end,
to the plastic, the paper,
and the little tin cans,

that floats in our rivers,
and into our bays,
out to our oceans,
and far, far away?

Glossary

Algae - a plant like organism which is usually photosynthetic and aquatic. Grows worldwide in both seawater and fresh water.

Chesapeake Bay - is the largest estuary in the United States. More than 150 rivers and streams from the District of Columbia, Maryland, Virginia, Delaware, New York, Pennsylvania, and West Virginia drain into the Chesapeake Bay.

Debris - trash, litter, garbage found both on our land and in our waters.

Fertilizers - added to our soil to help plants grow. Nutrients such as nitrogen (N) and phosphorus (P) are fertilizers. Too much of these can affect our water quality.

Menhaden - a filter feeder fish which can filter up to 4 gallons of water a minute, helping keep our waters clear. It lives on plankton that it catches. It also is a natural check to the harmful algae blooms that affect our Bay.

Oyster - is a bivalve mollusk which lives in a marine environment. Chesapeake Bay is one of the largest homes for oysters. Oysters consume algae by filtering water up to two gallons an hour, helping clarify our water.

Red Tide - also called a harmful algae bloom occurs when colonies of algae grow out of control. It can be harmful to fish, shellfish, marine mammals, birds, and even humans.

Rivers - some of the largest rivers that directly affect the Chesapeake Bay are; The Susquehanna, Patuxent, Potomac, Rappahannock, York, and the James rivers.

Acknowledgements;

It is our responsibility to protect and keep clean the Rivers, Bays, and Oceans, which were Created for us!

Tonya Leigh
"To a wonderful woman whose passion
and gentle guidance helped propel this project forward!"

Rick and Camie
You all are phenomenal. You brought the message
and ideas to life with your talents.
The illustrations are awesome!

Nancy L.
Where do I start? Over the years I've known you,
what a neat mentor and friend you are.
Thank you so much for the knowledge and time given.

To family and friends
Thank you for your patience & support.

To Cariese and Pete
The catalysts to the project, who helped bring this all together.

To the infinite possibilities when Imagination, Inspiration, and Illustration are combined.

Wonderful Information Sources

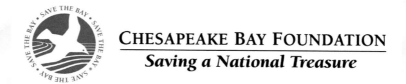

www.cbf.org

www.elizabethriver.org

www.longwood.edu/cleanva/

www.lrnow.org

Made in the USA
San Bernardino, CA
13 July 2017